Lucas Coelho de Avila

3D printing: a range of opportunities in product development

3D printing: a range of opportunities in product development

Lucas Coelho de Avila

3D printing: a range of opportunities in product development

3D printing and the possibility of creating specific products

ScienciaScripts

Imprint

Any brand names and product names mentioned in this book are subject to trademark, brand or patent protection and are trademarks or registered trademarks of their respective holders. The use of brand names, product names, common names, trade names, product descriptions etc. even without a particular marking in this work is in no way to be construed to mean that such names may be regarded as unrestricted in respect of trademark and brand protection legislation and could thus be used by anyone.

Cover image: www.ingimage.com

This book is a translation from the original published under ISBN 978-613-9-65607-3.

Publisher:
Sciencia Scripts
is a trademark of
Dodo Books Indian Ocean Ltd. and OmniScriptum S.R.L publishing group

120 High Road, East Finchley, London, N2 9ED, United Kingdom
Str. Armeneasca 28/1, office 1, Chisinau MD-2012, Republic of Moldova, Europe
Printed at: see last page
ISBN: 978-620-7-79198-9

SUMMARY

ACKNOWLEDGMENTS

I would like to thank everyone who helped me in any way and contributed to the completion of this work.

First of all, I would like to thank God for allowing me to get this far and for guiding me all the way through university.

To my father Ailton Letier de Avila and my mother Edma Coelho de Avila, for always supporting me and giving me all the support I needed to complete my studies.

To all my other family members who have always supported me and cheered me on, especially my sister Francisca Caroline Rangel de Avila and my brother Saulo Rangel de Avila, for all the encouragement and support they have given me since the beginning, when I decided to do this degree.

To my advisors Luiz Henrique Zeferino and Gudelia Morales, for their patience and help during the preparation of this TCC.

To every teacher I met during my degree and who shared some of their knowledge with me, thus contributing to my academic training.

To all the friends and colleagues I met during this course, even those whose contact was brief, but who somehow marked my career and gave me moments of pleasure and joy. In particular, I would like to mention my friend Juliana Damaris Candido, with whom I was very close and who was my companion during most of my academic work.

I would also like to thank UENF, and all the staff who work or have worked there, whose work has been fundamental to the realization of my academic history.

Finally, I would like to thank all the people who, although their names are not mentioned here, have helped me and contributed to the completion of this work.

SUMMARY

3D printers have come up with a variety of models in line with advances in research and technology. Among the many existing types, there are those that work with lasers, light beams, inkjets, liquid resins, among many others. For our case, we're going to focus on printers that work with plastic, based on extrusion. This is one of the simplest methods present in the cheapest machines, in which the extruder head releases subsequent layers of a heated plastic material, in order to provide the desired contours for the object, according to the design and specifications, which are dimensioned using software. Some of the software used is Blender, OpenSCAD, 3D Builder or even AutoCAD. The cost involved is low, in the case of small batch production, since only plastic is needed as a raw material, usually ABS (Acrylonitrile Butadiene Styrene) or PLA (Polyacrylic Acid); printing is usually quick and easy. However, despite all this, there are still major problems related to 3D printing, which can lead to higher costs, waste or deformations in the object. One example is the impossibility of interrupting production. In the event of an accident, such as a power cut, production cannot return to the starting point and has to be restarted. The city of Campos dos Goytacazes, for example, suffers from frequent power outages (most often caused by lightning), which can cause an interruption if the printer is running. Another problem concerns the difficulties of printing "floating" parts of an object, since the action of gravity would cause deformations in the part, so it is necessary to use supports while printing, which also causes waste. As can be seen, 3D printers offer a vast field of applications and studies, and can be related to various areas, such as the planning and design of new products, in which it is possible to research ways of "easing" their limitations, thus being able to take advantage of the advantages they offer more and more. The challenge is to find ways of optimizing the process, avoiding waste, reducing costs and improving the quality of your printed parts.

Keywords: Rapid prototyping, extrusion, 3D modeling, multiple uses.

CHAPTER 1

INTRODUCTION

In recent times, a new and promising type of technology has been developed and is gradually coming to the fore. This is 3D printing or rapid prototyping, which consists of printers capable of printing an object in three dimensions, as specified by the user.

According to Takagaki (2012) 3D printers are becoming popular at an impressive rate, due to the development of technologies that have made them accessible. This represents a major attraction for users looking for a fast, effective and low-cost production method for specific parts.

The concept of three-dimensional printing is one and the same, it always aims to produce a detailed object with volume and depth, however, even for a single application there are different technologies (PANKIEWCZ, 2009 cited in COSTA *et al.* 2015).

According to Magnus Herman, director of the company Feito Cubo Ltda., people usually use 3D printers to replace or reconstitute a defective object, or to build a model/prototype of a product that will later be manufactured on a large scale. However, it is not recommended to use 3D printing to manufacture products on a large scale, as it would not be economically viable (COSTA *et al.* 2015).

Rapid prototyping (RP) is a term that has been discussed by the manufacturing industry for many years. Defined in the mid-1980s, PR was used to describe a series of technologies capable of building product prototypes in the early stages of their development, in a rapid and automated manner. There is a wide variety of techniques that combine flat layers of materials arranged in sequence to form a solid three-dimensional object (INFORÇATTI NETO, 2013 cited in PALLAROLAS, 2013).

The 3D printer does the same job as the conventional printer, except that instead of simply putting ink on a sheet, it creates objects in 3 dimensions (BAIÂO, 2012).

According to Volpato et al. (2007), at that time most of this 3D technology was in the possession of large and medium-sized industries, but there was also a lot of dissemination of the technology in small and micro companies, and the trend is for its presence to become increasingly common in the market and among industries.

3D printing also aims to decentralize industries, facilitating production processes so that there is no need to waste months calculating the practical execution of a project; instead, everyone will be able to create their own objects as they imagine them. With the study of this new technology and the compactness and ease of operation of new 3D printing machines that will be developed in the future, it is estimated that people will adopt the equipment in their own homes (JUNIOR; MARQUES, 2013).

1.1 . General Objectives

In general terms, this work was designed to discuss 3D printing, presenting its characteristics, its history and evolution, how it works, how it can be used, its advantages and disadvantages and its possible contributions to various areas, such as engineering, medicine, architecture and education. There are also some interesting facts about this type of three-dimensional printing and an analysis of its future, based on the current market scenario and trends.

1.2 . Specific Objectives

The specific objective is to carry out a practical application of the use of 3D printing based on the product design methodology, thus demonstrating how 3D printers can contribute to the creation of innovative and differentiated products, on a smaller scale, or ergonomic products capable of helping workers carry out their routine activities. In addition, the advantages and disadvantages of buying a ready-made 3D printer or just buying the parts and assembling them will also be analyzed.

1.3 . Justification

The topic was chosen because 3D printing has great potential for development, with high chances of bringing benefits to various areas, such as architecture, engineering or medicine (especially the construction of prostheses for people who have been victims of accidents or birth problems), or even in teaching, via educational robotics.

The work was also inspired by the existing 3D printer at UENF, owned by Luiz Henrique Zeferino, the professor who supervised this project. The printer belongs to the GTMax3D Core A1 model, which works with extrusion-based plastic, and was responsible for helping with the applications by printing the planned products.

1.4 . Monograph Structure

This work consists of five chapters, the first of which is the introduction, objectives, justification and structure. In the second chapter, a literature review will be presented, which will cover the history of 3D printing, an overview of printers, their multiple uses, advantages, disadvantages, possible applications, trends and prospects, etc. Three types of 3D printer on the market and some of the main software used to build three-dimensional objects will be briefly presented. The third chapter will be dedicated to how the printers work, so it will cover how to assemble the printers, and will discuss the advantages and disadvantages for the user, between the options of buying a ready-made printer, or just buying the parts and assembling them. This chapter will also cover the main plastics used for three-dimensional printing and the Rep Rap assembly trends. In chapter 4, two practical applications of 3D printing in relation to the Product Design methodology will be presented, through the planning and design of two different products. Finally, in chapter 5, the conclusions and suggestions for future work will be discussed.

CHAPTER 2

LITERATURE REVIEW

2.1. 3D printers

There are currently several 3D printing technologies. All technologies are based on the principle of making several slices of the figure, usually horizontally, obtaining a thin layer that is printed through the process of depositing materials from the solid parts of the figure. By superimposing the various layers one on top of the other, we obtain the desired final object (TAKAGAKI, 2012).

3D printers are rapid prototyping machines developed with the aim of creating innovative products in the shortest possible time, differentiating them from conventional machines. At first, the machines were only used in industries, but the process has expanded, and now the researchers' main goal is for them to also be used in offices and private homes. The development of prototypes using 3D printing takes place in a similar way to ordinary printers, where the printhead deposits the ink on the paper, line by line. In the three-dimensional printing system, the product is developed graphically in 3D in computer software and then the model is converted into coordinates, divided into flat layers that are transferred to the printer in machine language. The construction material present in the printer head is deposited on a platform according to the final design, forming the prototype or desired product. The printing process uses plastic materials, resins, photo polymers and some specific metals, depending on the technology used (VOLPATO, *et al.* 2007).

The printing stage begins by preparing the STL extension file containing the object's 3D model, using software designed for this purpose. The aim of preparing the 3D model is to generate instructions so that the 3D printer can build the object. This preparation is called "slicing", since printing takes place layer by layer, i.e. the software reads the file containing the 3D model and divides the object's design into layers. It then generates a code that the 3D printer's processor can interpret so that the electronic components can be moved and the object built (AGUIAR, 2016).

According to Fernandes (2016) rapid prototyping is a technology on the rise, both in industries and in the commercial environment, so it is believed that it should become a very fast and efficient form of production.

Due to its ease and speed of production, the technology could become increasingly common in people's daily lives, transforming itself into a manufacturing tool, thus being called manufacturing or rapid production (VOLPATO *et al.* 2007).

2.2. History of 3D printers

3D printing technology originated fundamentally from two hitherto separate fields of study, topography and photo-sculpture, when Wyn Kelly Swainson proposed the direct manufacture of a part through the selective catalysis of a polymer at the intersection of two laser beams (BOURELL *et al.*, 2009 cited in MONTEIRO, 2015).

In the mid-1980s, Rapid Prototyping (RP) emerged, a term that defined a group of technologies that literally built prototypes at the initial stage of a product's development, materialized quickly and automatically (CAMPBELL et al., 2012 cited in PALLAROLAS, 2013).

According to the Antônio Barbosa organization (2012), laser printers were created in 1983 by Hewlett Packard and Canon. This device used a special laser for printing, which greatly improved the quality of printing in relation to ink, which in other printers would get wet and end up dripping, which didn't happen in these new models. The images and documents printed by these devices are of good to excellent quality (BAIÂO, 2012).

The first 3D printer was invented in 1984 in California by American Chuck Hull. At first they were expensive, large and difficult to handle.

The first commercial application of 3D printing technology was in 1987 by the company 3D Systems, which used the stereolithography system, *Stereolitografy* (SL), in which a photosensitive resin solidified when exposed to ultraviolet light (TAKAGAKI, 2012).

From 2012 to 2015, new printers and materials, such as carbon fiber and glass, allowed for more applications, which increased user demand. Segments such as the automotive, pharmaceutical, medical, gastronomic and architectural industries have already embraced the tool. According to the same consultancy, 5.6 million pieces of equipment are expected to be sold in 2019, an annual growth of 64.1% in the sale of 3D printers (SIGN and SILK SCREEN, 2015).

According to the renowned consultancy Gartner, 3D printing is a trend that is likely to grow immensely over the next few years, with sales and use in a wide range of sectors expected to increase. Although it's not such a new technology, it's gaining more and more relevance, as companies have started to use 3D printing to build important parts for their businesses (ÉPOCA NEGÓCIOS, 2017).

2.3. Programs used for modeling 3D drawings.

In order to be able to print an object on a 3D printer, you must have a 3D drawing in hand that shows in detail what the object will look like. If you don't know how to model objects in software, you can download ready-made models from the internet. However, if the user wants to print an object that meets a specific need, whose model doesn't exist on the internet, the only way is to learn how to model it in some 3D software. There are several types of 3D software, some of which we will introduce below.

2.3.1. Blender

It has complete resources for modeling and producing 3D objects, and a non-linear editor that allows you to build your own professional animations. Ideal for creating the most diverse types of games, it also supports various types of extensions, both 2D and 3D; its interface is simple and customizable, with a high number of customization possibilities, making it ideal for novice users. It allows you to save the modified object and export it to a different stl file, so that the user can configure the object to their liking before printing. Figure 1 shows the Blender interface.

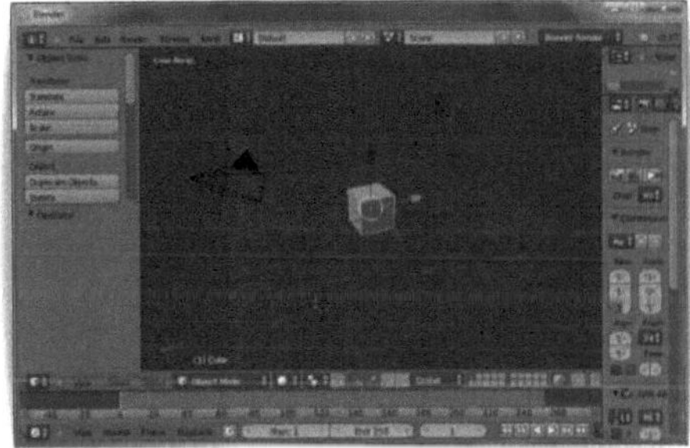

Figure 1: Blender interface

Source: BAIXAKI

2.3.2. Openscad

This is software for creating solid 3D objects. Unlike Blender, it doesn't focus on the artistic aspects of modeling, making it ideal for creating machine parts, but not very effective for creating animated films, for example. Its working mode is not interactive like that of most software, being a kind of compiler that reads the program file describing the object and creates the model from that file. However, although it provides the designer with fewer visual resources, it offers better control over the modeling, allowing any step in the process to be easily changed, thus enabling the designer to obtain a more precise drawing in terms of its shape and size configurations. It also allows the compiled model to be exported in a variety of formats that can be opened for future manipulation by other CAD applications. If the user doesn't know the language, there are ready-made examples on the web. Figure 2 shows the Openscad Interface.

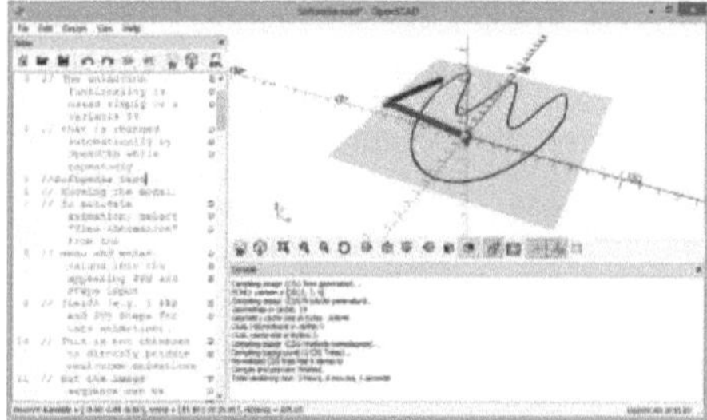

Figure 2: Openscad interface

Source: SOFTPEDIA

2.3.3. Autocad

It is the most widely used software in the fields of engineering, architecture and product design, and is one of the most complete and professional tools on the market. It has numerous functions that make it possible to create objects in three dimensions, even building plans for complex projects. Over time, the program has been modernized and made more accessible to beginners by presenting a more intuitive interface, which is also very useful for modeling 3D objects.

Represented by Figure 3:

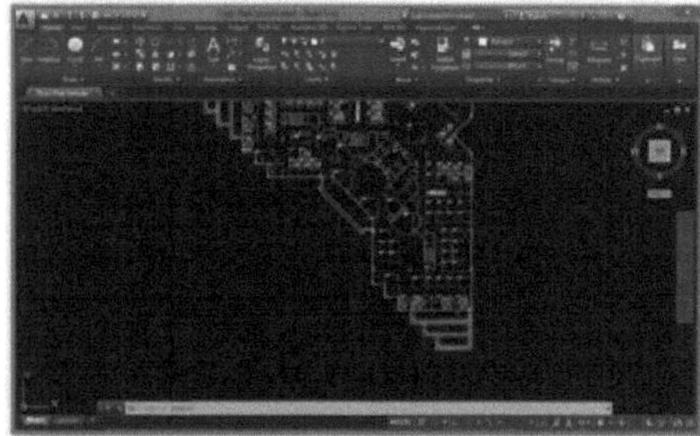

Figure 3: AutoCAD interface Source: BAIXAKI

2.4. Advantages and disadvantages of using 3D printers

3D printers have a great advantage in terms of cost and time if they are used to produce parts in small quantities. Their low cost is due to the fact that the material used, usually ABS or PLA plastic, is cheap.

It has the advantage of printing the object already in its final desired shape, eliminating the need for finishing, unlike machining, where a solid piece is produced, which is then subjected to the action of machines to acquire the desired contour, which causes the loss of part of its material.

In cases of low production needs, it is more efficient than plastic injection molding, where it is necessary to create a mold for the part. The manufacture of the mold generates a high level of work, costs and time, so if you only need to produce a small quantity of parts, the plastic injection method would be extremely disadvantageous. It should be noted, however, that for large-scale production, the plastic injection method is much more advantageous than 3D printing. This is due to the fact that, despite having a large initial fixed cost for manufacturing the mold, the variable cost for the plastic injection method is very small, so even for large production runs the total manufacturing cost is still very close to the fixed cost, with no significant variations. The opposite is true for 3D printing. The initial cost is low (almost negligible in the case of very small batches), but the variable cost is much higher, meaning that the total production cost of 3D printing quickly equals that of plastic injection, as the production quantity increases. Therefore, after a certain point, which varies depending on the size and complexity of the part, it becomes more viable to opt for plastic injection rather than 3D printing, as shown in Figure 4.

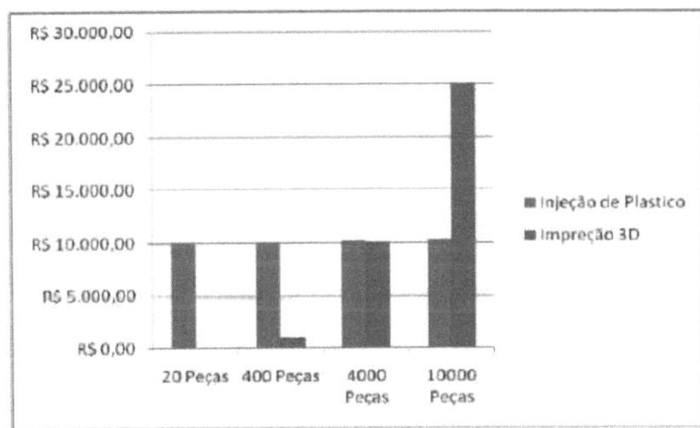

Figure 4: Production costs x quantity of parts

Source: MARIOTTO, 2013

As we can see, 3D printing only offers cost advantages in the case of small batches. For the manufacture of the part analyzed, from 400 pieces onwards, it is cheaper to produce using the plastic injection method.

Table 1 compares the main advantages and disadvantages of 3D printing:

Table 1: Advantages and Disadvantages of 3D Printing

Advantages	Disadvantages
Low production costs for specific parts.	High production costs for large quantities of parts.

It prints the object already in its final format, eliminating the need for finishing, as well as working with recycled material, making it a sustainable method.	Expensive printers, difficult to assemble and maintain.
It offers the possibility of a customized product.	It requires knowledge to model in software.
It has a wide range of applications in various fields.	It causes waste if there is an interruption in production.

Source: Prepared by the author

2.5. Multiple uses for 3D printers.

Taking into account the fact that they can produce a variety of objects, 3D printers can become useful in a number of areas. According to Table 2, we can highlight some of the possible contributions of 3D printers.

Table 2: Contribution of 3D printers to different areas

Area	Contribution
Art	Jewelry, sculptures
Entertainment	Figures, miniatures, toys
Architecture	Building visualizations
Health	Implants, prostheses, educational models.
Kitchens	Pasta and chocolate
Industry	Spare parts for machines.

Source: TEC MUNDO, 2013

Although some of these potential activities are yet to be explored, there are already reports of 3D printers being used to print pizzas in Barcelona. The printer is called Foodini and was created by the Catalan start-up Natural Machines. The object is presented like a large microwave oven, with a touch screen where the user selects the commands, from the shape of the dish to the ingredients. The food is then printed according to the shape and ingredients specified (MWC, 2016).

In China, too, there are reports of a car produced using 3D printing. Weighing 500 kg, the car (pictured in Figure 5) is powered by electricity and does not require the use of gasoline or alcohol. It took 5 days to print, far less than if it were produced in the normal way, which takes an average of around 30 days (GLOBO. COM, 2015).

Figure 5: Chinese car printed with a 3D printer Source: TEC MUNDO, 2015

According to Wohlers (2008) cited in Souza and Ulbrich (2009) the distribution of the use of rapid prototyping by the industry is consumer electronics (20.8%) followed by automotive (16.8%) and that with the evolution of technology this process is already used to produce the final part (RIBEIROS *et al.* 2013).

2.6. 3D printer examples

There are different types of 3D printers. Many of them have already arrived in Brazil. These include

2.6.1. Cliever CL - 1

It is the first 100% Brazilian 3D printer. It was created by Cliever Tecnologia, a company incubated at the Pontifical Catholic University of Rio Grande do Sul, and was launched to the world in June 2012. Made of steel and carbon and weighing around 12 kg, it uses plastic filaments as raw material to make 3D objects. It is shown in Figure 6:

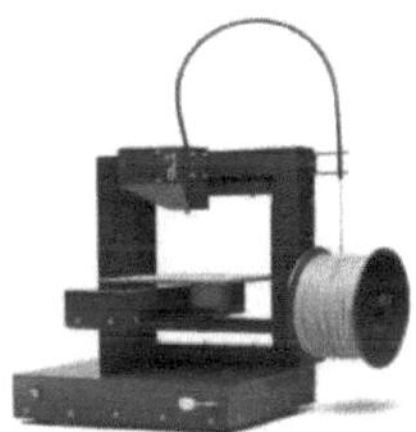

Figure 6: Cliever CL - 1 3D printer

Source: TECMUNDO, 2013

2.6.2. Makerbot Line

It is one of the best-known models in the world, and can also be purchased in Brazil with support in Portuguese. The printers in the Replicator 2X range use FDM (fused deposition) printing technology, usually with ABS, working at temperatures ranging from 15° to 32°C.

They weigh around 13 kg and come with software included. Figure 7 illustrates a printer from the MakerBot range:

Figure 7: MakerBot 3D printer

Source: BEEL, 2014

2.6.3. Cube Line

It prints on ABS and PLA materials and comes with a WiFi connection, allowing you to read files without being connected via cable to another device. It comes with software that allows you to create designs more quickly and with less technical knowledge, as it doesn't require the use of 3D modeling applications. It can be bought in Brazil for around R$6,690.00; its enhanced version (Cube X) works in a professional way and has three print resolution modes and can be bought for around R$11,000.00. Figure 8 shows a Cube printer:

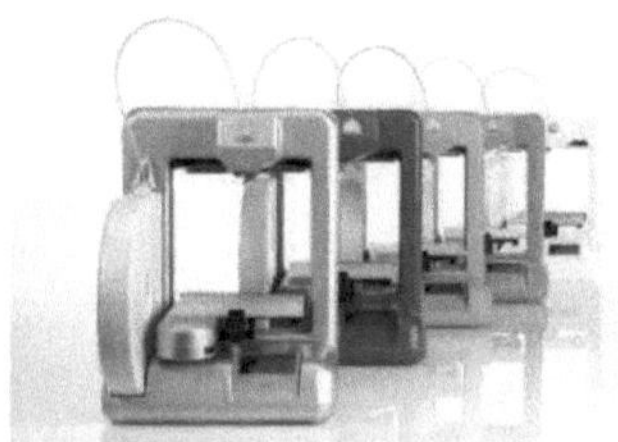

Figure 8: Cube Line 3D Printer Source: TECMUNDO, 2013

2.7. Main problems faced by 3D printers

Despite representing a major scientific and technological advance, 3D printers face some problems and still have a lot to improve if their use is to be considered truly efficient. There are some problems that can make their use unfeasible and cause waste.

One of the problems is the impossibility of interrupting production, because if there is a power failure, production cannot return to the starting point and will have to be restarted, thus losing the material that has been used up.

Another problem concerns the difficulties of printing the "floating" parts of an object: 3D printers print objects from their base, and then deposit layers of plastic on this base to form the desired object. However, when an object has an extra part to its base, which is already at a certain height, this part is called the "floating part" and will suffer the impacts of gravity, since there is no support base for this part of the piece. In this case, it is necessary to use supports or rods to hold it up while printing. The supports are usually made from biodegradable plastic and are removed after the print is finished, thus generating material losses and extra costs. In addition, complex geometric shapes can take a long time to print and still end up lacking in detail. Apart from that, there are other problems, such as the fact that printing is not as easy as it looks, and requires various types of technical knowledge and skills on the part of the user, both to set up the printer and to carry out maintenance, 3D modeling, etc.

One of the solutions suggested revolves around printing the parts separately and then joining them together. For example, in the case of a man with his arms outstretched, it would be enough to print the arms separately and then fit them to the body. In the case of a pyramid or cone, for example, it would be enough to print the base instead of the apex, which would be absolutely fine. However, it would not be possible to adopt these strategies in the case of some more complex geometric shapes, which makes 3D printing unfeasible for these parts, as it would necessarily have to cause some waste.

2.7.1. Negative expectations regarding the growth of 3D printing

Despite its high growth expectations, there are experts who fear that the consumer goods industry will suffer if 3D printing grows too much. This is due to the fact that if people have the possibility of producing a product in their own way, easily and quickly, and in their own home, they will no longer need to buy them, which will lead to the end of several industries. In addition, the inclusion of 3D printing in many different areas could replace human labor, reducing the need for manpower and generating unemployment. The market will therefore have to adapt to these changes if growth is to occur as expected.

Another very worrying problem is the use of 3D printers for illicit or immoral purposes, in order to provide resources for robbers, traffickers or people of bad faith to act.

Burglars could use digital manufacturing to break into homes. It would be possible, from a single high-resolution photo of the keys to the potential victim's home, to print an exact replica. Art. 150 of the Penal Code will soon be rewritten to cover this possibility. And what about drug trafficking? If there are already devices capable of printing painkillers and other drugs on demand, it won't be long before traffickers replicate methamphetamine, crack and codeine, leveraging the market worldwide (SOUZA, 2016).

According to Souza (2016), another danger is the theft of intellectual property, as offenders can use 3D printing to steal intellectual property, such as Gucci bags and Rolex watches, which can easily be scanned at very high resolution, producing copies that are as visually perfect as the original products. The Gartner group estimates that 3D printing will result in the loss of at least US$100 billion annually in intellectual property across the globe.

2.8. Trends and possible applications of 3D printing

Bone lesions that cause deformations to the craniofacial anatomy are usually complex problems to solve, due to the aesthetics and functions involved. The anatomical complexity of the area requires high levels of adaptation of any system used. The conventional method used to reconstruct these defects is based on bone grafting, which is able to cover and complete the defects (JALBERT, 2014 cited in MORAES, 2015).

With advances in technology, it is possible to 3D print metal shapes of the desired geometry using Selective Laser Melting technology in the healthcare sector. Selective Laser Melting Technology (SLM) uses powerful 3D printing technology to shape the desired metal geometry by melting metal powder layer by layer. The metal used to create the customized implant is titanium Ti - 6% Al - 4%.% by weight (grade Ti64 23) with a low oxygen content (MORAES, 2015).

With the integration of RP, computed tomography scans can be taken and, consequently, 3D printing is able to print the physical model on another material, making it easier to analyze the region (JUNIOR; MARQUES, 2013).

With this, we can see that 3D printers can be great allies when it comes to problems related to anomalies in parts of the body, such as the absence of an organ or compromised functions.

On 06/03/2016, Fantàstico (a Brazilian television program broadcast on Sundays by Rede Globo) aired the series Fab Lab: Do it yourself, where a challenge was solved consisting of creating a prosthesis on a 3D printer for an 8-year-old boy who was born without four fingers on his left hand. The prosthesis was created in a short space of time and received public input on ideas and solutions for the child, as well as being a lighter and cheaper option than the conventional one. This is due to the fact that a prosthesis, because it is a part designed exclusively for a person's needs, is most advantageously produced on a 3D printer, thus dispensing with the unnecessary creation of a mold, which would be much more expensive to use just once.

In addition, Florêncio *et al.* (2016) highlight the possibility of using 3D printing in relation to the construction process. According to the authors, the process would consist of pumping concrete stored in a reservoir through an extruder nozzle coupled to computer-controlled shafts with xyz movement, very similar to the process carried out by printers using polymers. This would provide a number of benefits, such as reduced costs and construction time, greater precision in the realization of projects, a reduction in environmental impacts, as well as logistical benefits, since printing the buildings directly on the ground would reduce the transportation of parts.

According to Khoshnevis (2012) quoted in Florêncio *et al.* (2016), "by scaling up conventional 3D printers, entire neighborhoods of decent housing can be built, using a fraction of the cost and time in a much safer environment and offering unprecedented formal flexibility".

2.9. 3D Printing and Agile Engineering

The agile movement seeks to improve the development and design of new products by adopting practices such as incremental development, iterative development, iterations, combined with internal customer perspectives and intrinsic risks. It differs from traditional models in that it carries out several tasks and stages of production at the same time and integrates them with the others as soon as it is complete, whereas in traditional models one stage only begins when the other has been completed (NETO, 2016).

The aim of Agile Engineering is to bring together the various areas related to the product development and production process, as shown in Figure 9:

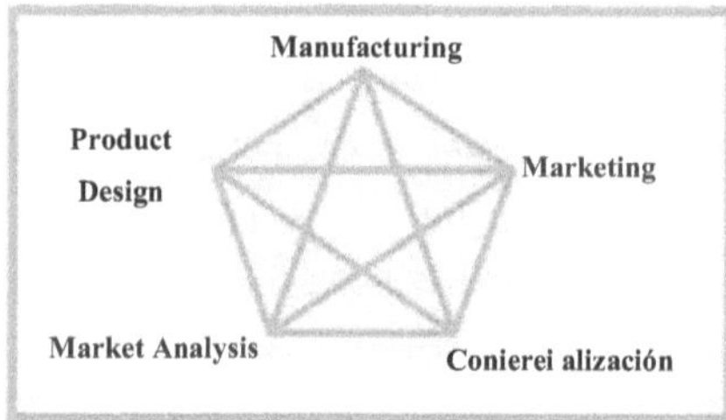

Figure 9: Concurrent Engineering Model

Source: VASCONCELOS 2001

3D printers provide a very favorable environment for the practice of Agile Engineering, because if an author of the process has an idea, they can print a prototype and then defend it. This avoids the need for multiple meetings, which delay development and hinder the company's productivity. They are therefore important for Agile Engineering and thus contribute to the design of new products.

2.10. 3D Printing: Environmental and Social Issues

3D printers can also bring a number of environmental and social benefits: the production method reduces waste, as it avoids the formation of a mold that will then be discarded, and it works with recyclable material, which, being sustainable, can bring important benefits in terms of the quality of life of waste pickers.

The waste pickers, as can be seen in Figure 10, are mostly adults between the ages of 30 and 49:

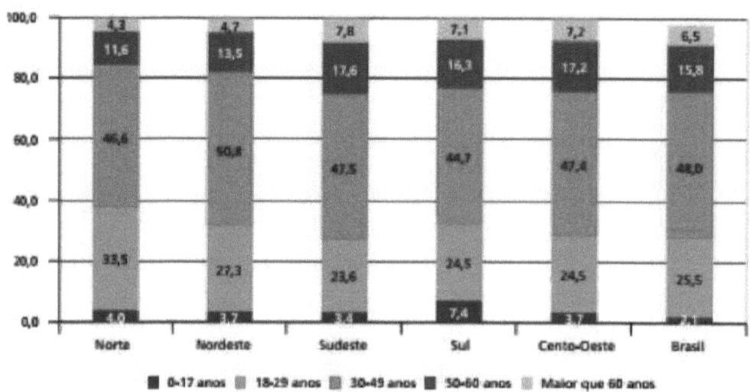

Figure 10: Waste pickers by age group

Source: IBGE, 2012.a

These are workers who depend on the collection of recyclable materials for their jobs and livelihoods. Little money, a lack of adequate housing and poor working and hygiene conditions are the reality of their lives.

The work done by these workers consists of collecting, sorting, transporting, packaging and sometimes processing solid waste with market value for reuse or recycling. By giving value to waste through their work, waste pickers "end up renaming it, feeding the very process of positive re-signification of their work activity" (Benvindo, 2010, p. 71).

The transformation of these materials into new goods and their reinsertion into the production cycle generates "positive benefits for nature and society, as it saves natural resources and space for storing waste" (Magalhâes, 2012, p. 14).

ABS plastic-based 3D printers can help improve the lives of garbage collectors, because they use recyclable material for their raw materials, which can increase the price and sales of this recyclable material, so that the work of these collectors is more valued, allowing them to make more profit from their sales (since they currently earn almost nothing).

CHAPTER 3

HOW 3D PRINTERS WORK

3.1. How does a 3D printer work?

To assemble a 3D printer, you need to know its toolset: electronic software, firmware, control and cutting software (EVANS, 2012).

If we start at the bottom of the chain, we have the electronic platform, which controls the 3D printer and communicates via its firmware to the control application. The electronics include various parts that work together to build the prints. These components include a micro controller, a main board, motor controllers, stepper motors, a hot end, limit switches and temperature sensors (EVANS, 2012).

In Figure 11 you can see how the different parts of the electronic system are interconnected, with the arrows representing the direction of control from one component to the next.

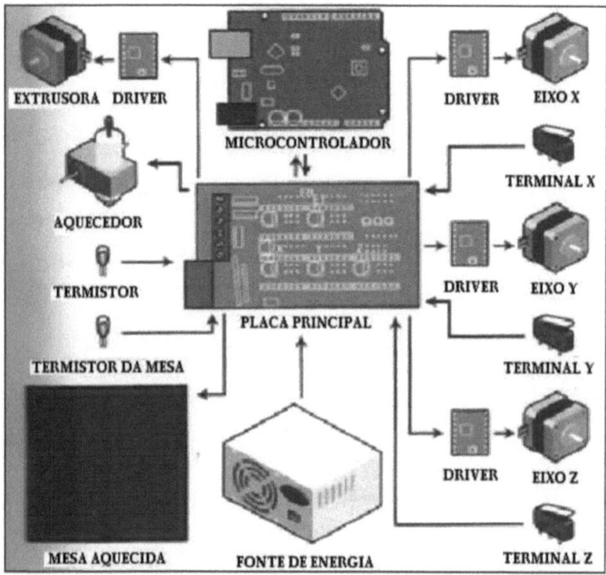

Figure 11: Map of the electronic parts of a 3D printer

Source: Adapted from EVANS, 2012

The microcontroller, found on the main controller or on a separate board entirely, is a small, simple computer that executes specialized code called firmware to allow it to read and interpret sensors, such as temperature and limit switches, so that it can then act on its actuators, such as motors (EVANS, 2012).

The small board is a simple chip board. It can drive a single stepper motor and is used by some controller boards to drive the printer axes. The power supply provides energy for all the electronic components to work. The arduino acts as the brains of the platform, while the shield provides the switching hardware for the heater and the print table (EVANS, 2012).

The firmware serves to control the specialized codes loaded into the microcontroller, and is responsible for interpreting the code commands sent to the electronics from the printer's control application.

The better the firmware does its job, the more efficient the printing will be (EVANS, 2012).

In order for the path to the printer's extruder to be generated, it is necessary to use an application called *Slicer* so that the 3D model can be sliced into suitable layers. This process causes the code to tell the 3D printer where to move the extruder, and then the amount of plastic needed for extrusion. These commands are sent from the printer's control software to the firmware, which will interpret these codes in order to control the printer's motors and heaters (EVANS, 2012).

The controller is the 3D printer's Interface, where the whole chain comes together. From it you can connect the printer to its firmware, move the different axes, as well as read and set the temperature for printing (EVANS, 2012).

3.2. Assembling the Parts

The first step in assembling your 3D printer is to test its components separately. The controller, for example, can be tested by connecting it to the computer and loading the firmware. You should then run tests such as moving the axes (to check the rotation of the motors), checking the hot end and the LCD panel. This process is known as bench testing and seeks to ensure that the operations occur in the most appropriate way possible (BELL, 2014).

You should avoid cutting the wires, preferably bundling them with zip ties. The wires used must be special, designed for use in heater circuits, given the strong heating that the extruder and the heating plate can undergo. Some LCD panel components are sensitive to radiation from other wires, so you should make sure that the LCD cables are routed away from wires carrying power. A good method is to wrap the LCD cables in electrostatic discharge shielding. You should also use markers on the cable so that you can identify them in the future (Figures 12 and 13).

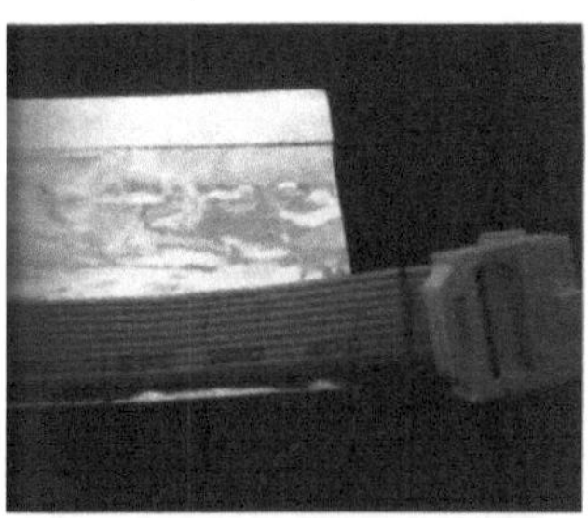

Figure 12: LCD cables before winding

Source: BELL, 2014

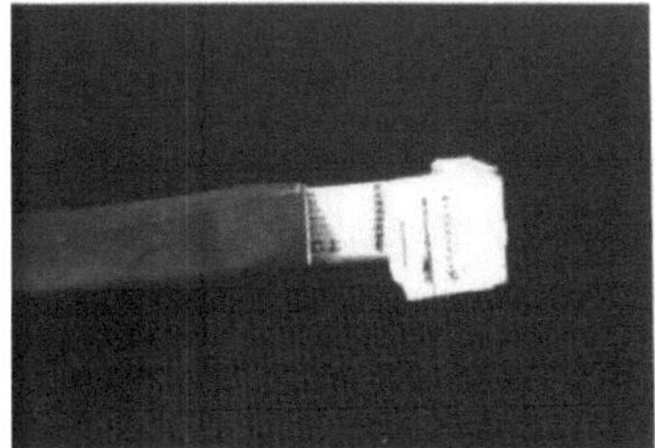

Figure 13: LCD cables after winding

Source: BELL, 2014

If your printer kit doesn't have a self-contained power supply, care should be taken about how you connect it to your home power supply, and the AC power connector should be deliberately protected. A popular method is to use an AC plug together with a switch and mount it to the printer in some way, either as part of a cover on the power supply or as a separate assembly. The best way to insulate the wires is to use heat pipes to cover the ends of the connectors, ensuring that all the high voltage wires are covered (Figures 14 and 15).

Figure 14: High-voltage wires before covering

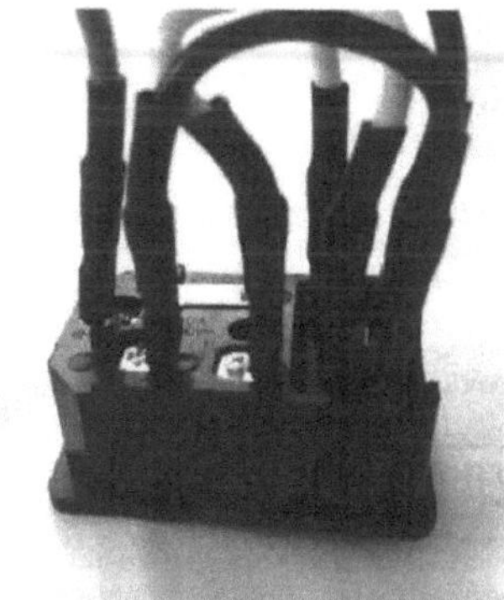

Figure 15: High-voltage wires after being covered

Source: BELL, 2014

Cables that connect to moving parts, such as the print bed, extruder and some shafts, are often very susceptible to failure due to the rapid movement to which they are subjected. For these areas, a stress relief mechanism should be used to keep the bending wires at a single point. Plastic wire wraps can be used around the bundled cables. Figure 16 shows the X-axis with a stress-relieving plastic wrap, one end of which is connected to the X-bracket and the other end to the structure, so the cable is not flexed in a single position.

Figure 16: Stress relief on extruder cables

Source: BELL, 2014

The cold end of the extruder is separated from the hot end and connected with a Teflon tube. If the user is using their own power supply, a different branch of the power supply should be used for each main component. Two conductors should preferably be used for the RAMPS (main board) in order to better distribute the load (BELL, 2014).

From there, the so-called "smoke test" must be carried out. To do this, connect the power supply and turn on the printer. If the firmware is loaded in the printer, the menus will appear on the LCD panel. If any smoke appears or any strange noises are detected, you should immediately cut off the power supply and investigate, but if there is no sign of anything strange, you can continue with the check. The next step is to check whether the terminals are set higher than the absolute minimum. If they are, this will allow you to make some adjustments, since the printer is already properly calibrated. However, if they are not, you should loosen the end supports for the X and Y axes, and move them towards the center of this axis (about 5 mm to 10 mm). In the case of the Z axis, you can adjust the limit switch to the maximum. If you have a fixed end of the Z axis, you can loosen the bracket and raise it about 5 mm (BELL, 2014).

Now you have to move on to the part that is generally considered to be the most fun: starting the printer driver software, inserting the USB cable into the computer, and connecting the other end to the printer. By pressing the connect button, the software will finally communicate with the printer. If you can't connect the software, you should check the port conditions of the printer software and try again. If problems persist, check the communication speed (some firmware versions allow you to set the communication speed, so look in the documentation for the correct speed). The standard three-pin terminals are represented by the letters NO, NC and C, which stand for normally open, normally closed and closed respectively. The correct wiring for each terminal must be checked with the supplier. The endstop, on the other hand, must be checked with a special G-CODE, the necessary code being M119, which must be issued via the printer driver software (Figure 17).

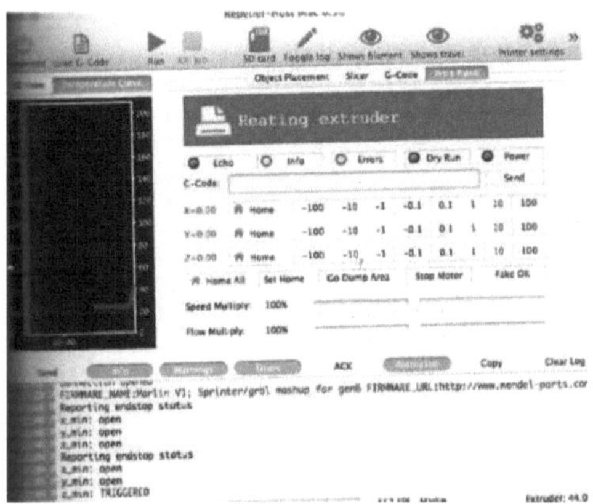

Figure 17: Checking the terminals with code M119

Source: BELL, 2014

Figure 17 shows that the firmware has returned the status of each end of the line. The first time all the terminals were marked open, but the second time terminal Z is marked as TRIGGERED. This is due to the top end being pressed manually. Sometimes, when using Marlin or similar firmware, it may be necessary to change the configuration to match the end-of-line behavior. If normal terminals are used, you don't need to adjust the firmware,

To check this, try moving the axes, one at a time, about 10 mm towards the positive side. For the X axis the positive side is to the right, for the Y axis it is backwards and for the Z axis it is upwards. Then reconnect all the axes and turn on the printer. Use the LCD panel or the printer controller software to move each 10 mm axis in the positive direction. If any of the axes do not move in the correct direction, you can correct the problem in the following ways:

•	If the connectors are not polarized, you can reverse the stepper motor connector on the main board, which will effectively reverse the direction of the motor.

•	If the connectors are polarized, you must change the direction in Firmware or change the wires on the motor (BELL, 2014).

There are also some tools that can be useful for building, maintaining and improving printers. These are

Wire strippers: there are two types, those whose insulation is removed manually and those whose insulation is removed automatically. The first type is more common, mainly for the majority of small jobs, such as repairing a broken wire, but the second, on the other hand, does larger jobs, such as bare wire 3D printer electronics wiring, without prefabricated connectors.

Soldering iron: necessary for most printers, especially those without pre-made electronics and cables, as it allows you to solder two wires together. Professional models have features that allow you to set the temperature of the wand, have various tips available, and tend to last longer.

Multimeter: used to measure electrical quantities, incorporating various measuring instruments. It is needed to carry out electrical repairs on the printer (BELL, 2014).

CHECKING THE HEATERS

To test the heaters, the hot end needs to be connected to the heated bed from the printer driver software. The hot end should be set to 200° C and the heated end to 100° C. You can use a temperature gauge to ensure that they are heating up. Some ends may give off a little smoke when first heated. This is normal, and using a fire extinguisher will only get in the way. However, if there is thick smoke or cracking sounds, you should disconnect the printer immediately and check the wiring for possible electronic faults (BELL, 2014).

EXTRUDER TEST

The last test to be carried out is the extruder test. For example, in the case of spring-loaded extruders, you press down on a plug and insert the filament, passing a drive gear to the hot end. Once heated to the appropriate temperature, use the printer's controller software to extrude a small amount of filament. If the extrusion goes smoothly, the printer is ready and all that remains is to calibrate it before printing the first object (BELL, 2014).

SOME ESSENTIAL PRECAUTIONS:

•   Never disconnect any components, especially stepper motors, while the printer is switched on. The printer must always be switched off before working with the connections.

•   Don't try to move the axes as much as possible. This value will be set as part of the configuration once the terminals have been set correctly. The printer can be damaged if it is moving faster than it should (BELL, 2014).

SOME TIPS:

•   The direction in the firmware can be changed.

•   If you are building a RepRap printer, the firmware (unless it has been changed) assumes that the rotation axis is in the O position.

3.3. Differences between buying a ready-made 3D printer or assembling it

The construction of a 3D printer encompasses the entire REPRAP class of printers, as well as any type that requires assembly skills. There are several reasons why users choose to buy the parts separately and assemble them themselves. As a first example, we can cite the savings factor, as it is cheaper to buy the parts separately than to buy them ready-made. Another benefit is the possibility of assembling it according to your preferences. By buying a ready-made printer, the user loses the opportunity to choose, having to use the product as it was offered. By buying the parts separately, the consumer has the flexibility to decide which brand, type or design each part of the printer will be. Another positive point is that by assembling the printer, the user acquires important knowledge and skills, which will allow them to identify and fix any defects, or improve it in the future by adding new parts or functionalities.

However, there are also some advantages and practicalities to buying your printer ready-made. One of the biggest advantages is time. It usually takes an average of 40 to 60 hours to assemble a 3D printer, so if the user is in a hurry to start printing, it would take a long time to learn how to build and then assemble it. Another reason is quality, especially in the case of professional printers, which tend to be of higher quality when bought ready-made. Finally, it is believed that the main advantage of buying the printer ready-made is the technical support, because if the user tries to assemble it without having the necessary technical skills and knowledge, this will cause them a loss in relation to the money they have invested buying the parts, as well as the fact that they will not have support from the supplier to solve future problems that the printer may present.

Table 3 summarizes the advantages and disadvantages of buying parts from a 3D printer and assembling them.

Table 3: Advantages and disadvantages of buying parts and assembling a printer

Advantages	Disadvantages
Saving money	It takes time to assemble, which gets in the way if the user is in a hurry.
Possibility of assembling the printer according to the user's personal taste or preferences.	It can cause damage to the user if he buys the parts and then realizes that he doesn't have the necessary skills to assemble them.
If you manage to assemble it, you will acquire knowledge that will enable you to repair or improve the printer in the future.	Printers tend to be of lower quality than those that come factory-assembled.
After assembly, the user is satisfied with the printer and the knowledge they have acquired.	Lack of technical support from the supplier if the printer has a problem and the user doesn't know how to solve it.

Source: Prepared by the author

We can therefore conclude that when making the decision between buying the printer ready-made or buying the parts and assembling it, the user must take into account whether they have the necessary skills and whether they have enough time to assemble it. In addition, there may also be secondary factors that influence the decision, such as the customer's propensity to risk, since when trying to assemble the printer, even if there is time and skill available, there is still the possibility of a problem occurring, and without support, this will cause damage.

3.4. RepRap printer assembly trends

The RepRap brand is the first major technological evolution of the most common 3D printers and was created by Adrian Bowyer (professor of mechanical engineering at the University of Bath in England) and his team. It can be used under the GNU General Public License, which considerably reduces the price of a 3D printing machine (REPRAP.PT).

RepRap is an open source desktop 3D printer that can print plastic objects. Most of RepRap's parts are made of plastic and RepRap itself can print these parts, so RepRap is a self-replicating machine, i.e. something that anyone can build, all it takes is time and the necessary materials. This means that with a RepRap, you can print dozens of sets of parts in a practical way, and you can even print another printer for a friend (REP RAP WIKI).

In other words, the RepRap movement refers to any technology capable of self-replication, such as 3D printers that can "print" their own parts, thus making it possible to build a new printer. This trend is helping to develop a greater number of printers, as well as making the technology cheaper and more accessible, as it allows complex objects to be manufactured without the need for an expensive and complex industrial infrastructure. There are still some components of the printers that are not replicable, such as sensors or micro controllers, but the trend is that with the passage of time and evolution, a 100% replication approach will be achieved.

3.5. Most commonly used materials for 3D printing

The most commonly used materials for 3D printing are ABS and PLA thermoplastics. We'll look at the main characteristics of each one separately and then make a comparison of which one is most suitable for use in certain situations.

3.5.1. ABS

ABS (Acrylonitrile Butadiene Styrene) is a thermoplastic derived from petroleum that has great rigidity, good resistance to impacts and high temperatures, and slight flexibility (compared to PLA). It offers advantages when printing strong, durable parts that are resistant to friction, impacts and high temperatures; however, it has the disadvantage that it cannot print the dimensions of the part precisely, and variations can occur between the object produced and the one modeled in the software.

It is a very easy material to finish, and can be sanded and machined easily. It is also soluble in acetone and can have its surface smoothed when dipped quickly in the solution (IMPRESSÂO 3D FACIL, 2015).

From an environmental point of view, it has advantages because it is recyclable, but it takes a long time to decompose and comes from oil, a non-renewable material.

3.5.2. PLA

PLA (polylactic acid) is a thermoplastic derived from renewable sources such as corn starch, cassava or sugar cane roots. It produces objects that are quite rigid and not very flexible. Due to its high hardness, it is less resistant to impacts than ABS, but it produces parts with more precise dimensions, so the final object will be more faithful to the one modeled in the software.

This material can be sanded and machined, but with some difficulty due to the heat generated in these processes, as it becomes somewhat pasty when heated by the friction of sandpaper and drills, and also due to its high hardness. The piece can deform, for example, when left in a closed car in the sun (3D PRINTING FACIL, 2015).

From an environmental point of view, it has advantages because it is a recyclable material, comes from renewable sources and takes little time to decompose.

3.5.3. Comparison between ABS and PLA

Table 4 provides a clearer and more summarized comparison of the use of ABS and PLA plastics in relation to certain criteria.

Table 4: Comparison between ABS and PLA in relation to specific criteria

Plastic	ABS	PLA
Sustainability	Worse	Better
Ease of finishing, e.g. sanding.	Better	Worse
Manufacture of fittings or assemblies	Better	Worse
More precise dimensions	Worse	Better
Impact resistance	Better	Worse
Resistance to high temperatures	Better	Worse
Production of pieces rich in detail	Worse	Better
Costs	Worse	Better

Source: Prepared by the author

We can therefore conclude that ABS is suitable for:

- Functional prototypes,

- Parts that need to be more resistant, whether to impact or temperature,

- Parts that need to be slightly flexible to fit together

- Parts for which the post-processing/finishing process is to be made easier (IMPRESSÂO 3D FACIL, 2015).

In other cases, where there is no need for any of the factors mentioned above, it is more feasible to use PLA, as it is more sustainable and less expensive.

CHAPTER 4

3D PRINTING AND PRODUCT DESIGN

Product planning and development is quite complex and multidisciplinary, as it involves several stages, such as planning, technical and economic feasibility studies, as well as market research, in an attempt to assess whether the product will be well accepted by consumers. The new product must meet various requirements, such as being aimed at a specific target audience, meeting quality requirements, functioning as expected, presenting an acceptable cost, being as sustainable as possible and containing aspects that differentiate it from its competitors. For this reason, it is extremely important to plan and develop products with a view to meeting all these requirements, so that the product succeeds and generates profits, and does not have to rely on chance (BAXTER, 2000).

According to Chiavenato (2005), product development is the area that deals with all studies and research into the creation, adaptation, improvement and enhancement of the products produced by the company.

One of the well-known factors in product development is the degree of uncertainty, which is very high at the beginning of the process and decreases over time; but it is precisely at the beginning that the greatest number of constructive solutions are selected. Decisions between alternatives at the start of the development cycle are responsible for 85% of the cost of the final product. The cost of modification increases throughout the development cycle, because with each change, a greater number of decisions already made can be invalidated (ROZENFELD *et al., 2006* cited in JUNIOR; MARQUES, 2013).

According to Freixo and Toledo (2003), making changes when the product is no more than a concept or idea is less laborious and involves fewer resources than making changes when the projects and processes have already been defined.

New successful methods have been created for product development, structured in stages that promote integration with other areas and execution planning. One of the most important integrations of the Product Development Process (PDP) is with rapid prototyping, a technology capable of developing functional components, prototypes and objects in the shortest possible time, in a way that respects the limits of the products, by means of 3D printing (VOLPATO *et al.,* 2007).

Integrating the concept into product development follows the guidelines of concurrent engineering and simultaneous engineering, which recognize the high cost of producing and developing products, through market analysis, trade and integration with product design. The aim is to bring these areas together in the same production cycle (JUNIOR; MARQUES, 2013).

Rapid prototyping makes it possible to manufacture parts or models of objects simply and quickly, making it possible to obtain assemblies or parts in quantities and conditions that cannot be achieved using the normal production line. It can be used to build study models, or simulations to test parts or new products before putting them on the production line, making it possible to notice and correct faults more cheaply and effectively (LEMOS, 2013 cited in LIMA; CORRÊA, 2016).

Therefore, it is possible to highlight several advantages and benefits that 3D printing can bring to the process of creating and designing a new product. According to Garcia (2010), the main advantages are a reduction in manufacturing time (because the process is carried out in a single step) and a reduction in costs, since it is possible to obtain quality prototypes at an early stage and at a low cost, which allows tests and trials to be carried out, preventing failures from occurring. In addition, it is possible to obtain parts with very complex geometries, which would not be possible with other methods.

4.1. Planning products to be produced using 3D printing

3D printers also make it possible to independently create products that can satisfy personal needs, such as tools, screws, remote control covers, kitchen utensils, or even utensils that don't exist yet or haven't even been thought of. Therefore, they offer the user the opportunity to undertake and create innovative objects, with particular characteristics, enabling new and differentiated products to be created.

Two practical applications of 3D printing will be presented below, where the steps involved in planning and making two products will be described in detail. The aim of these applications is to demonstrate how 3D printing offers the chance to design new products combining concepts such as sustainability (by producing using recyclable material), innovation (by allowing the possibility of creating improved and differentiated products) and creativity (by allowing the user to use their imagination and characterize the object in a way that satisfies their desires).

4.1.1. Characterizing the first object

The first object to be produced is a dice, a fairly simple object used in various types of games, whose primary target audience is children. The dice to be developed has some attractive features that set it apart from the rest: it can be used as a pencil holder, a decorative object, an element for games and can be turned into a keyring. The object will be modeled in Openscad software, a language that will allow it to be reproduced by a 3D printer.

The dice will consist of a solid cube with a 2.4 cm (or 24 mm) edge, which will have holes or cylindrical holes in its faces, signifying the value drawn. For the face representing the number 1, there will be one hole, for face 2 there will be two holes and so on. A very important detail to note is that the volume taken from each face to make the holes should be as close as possible to each other. This is strictly necessary so that you don't make a flawed die that will tend to fall with the larger volume face always facing downwards, a conceptual specification that is respected. In this way, it can be concluded, for example, that the volume of the hole in face 1 must be equal to the sum of the volumes of all the holes in face 6.

The die was intended to be used as a pencil holder on face 1, due to the fact that this is the face whose hole diameter is closest to that of a pencil or pen. Conventionally, the value of 7.7 mm for the diameter (which means 3.85 for the radius) and 8.0 mm for the height of the hole in face 1 have been proposed, so that the die can fulfill its function as a pencil holder or decorative pencil holder.

The volume of the holes can be calculated as the product of the volume of the cylinders by the number of holes in each face, using the equation: $V = \pi * r^2 * h *$ number of holes.

Where r is the radius and h is the height of the hole. The volume consumed for the hole in face 1 was then calculated:

Volume 1 $= \pi * r^2 * h *$ number of holes $= \pi * 3.85^2 * 8 * i = 372.53 \text{ mm}^3$.

The values of the radii and heights of the other holes remain unknown. So, based on the volume removed to make the hole in face 1, the values for the radius and height of the holes in the other faces were proposed, so that the volumes removed were equivalent to each other, as well as maintaining good aesthetics in relation to the final object. The proposed values for the radius and height of the holes for each face are shown in Table 5:

Table 5: Radius and height of each hole in millimeters

Face	Lightning	Height
1	3,85	8,0
2	3,4	5,1
3	2,45	6,6

4	2,6	4,4
5	2,4	4,1
6	2,35	3,6

Source: Prepared by the author

The volumes removed from the other faces were then calculated using the formula: $V = \pi * r^2 * h *$ number of holes.

Volume 2: $\pi * r^2 * h *$ number of holes $= \pi * 3.4^2 * 5.1 * 2 = 370.43 \, mm^3$

Volume 3: $\pi * r^2 * h *$ number of holes $= \pi * 2.45^2 * 6.6 * 3 = 373.38 \, mm^3$

Volume 4: $\pi * r^2 * h *$ number of holes $= \pi * 2.6^2 * 4.4 * 4 = 373.77 \, mm^3$

Volume 5: $\pi * r^2 * h *$ number of holes $= \pi * 2.4^2 * 4.1 * 5 = 370,96 \, mm^3$

Volume 6: $\pi * r^2 * h *$ number of holes $= \pi * 2.35^2 * 3.6 * 6 = 374.75 \, mm^3$

As we can see, the volumes spent on the holes for each face are very close to each other. The smallest volume was 370.43 mm^3 for face number 2, while the largest was 374.75 mm3 for face number 6. THE difference between the largest and smallest volumes is only 4.32 mm3, which represents a relative error of 1.2% in relation to the volume of the hole in face 1, this error being caused by rounding off the decimal places. In order for the volumes taken from each face to be exactly the same, numbers with several decimal places would be needed, which would make it difficult for the printer to be precise, so we decided to leave the values as they are, with a maximum of two decimal places. Furthermore, this relative difference in volume between the faces will not influence the honesty of the data.

4.1.2. Programming the object in Openscad

Once the dimensions of the edges, heights and radii of each hole have been defined, all the planning will be translated into the *Openscad* programming language so that the desired object can be modeled and printed.

On the website www.thingverse.com.br, you can find ready-made object models with the code normally available in the *Openscad* language. You then have to download the code for any data and make the necessary changes so that it meets the planned specifications.

Using this code, a window is generated on the *Openscad* screen with an image that allows a visualization of what the printed object will look like. It's worth noting that the colors present are only for better visualization, and don't mean that the object should necessarily be printed in that color, as it will be printed according to the color of the wire placed in the printer. The image is shown in Figure 18:

Figure 18: Data representation on the Openscad screen

Source: Prepared by the author

4.1.3. Adding the keyring function

It was planned to make a cylindrical part with one end slightly larger than the other. The larger end will fit into the central hole of the face with three holes, so the diameter of this piece will be slightly smaller than the diameter of the holes in face three. The end of the piece with the smallest diameter will have a small hole through it, into which a ring can be fitted, allowing this piece to be used as a keyring or decorative object.

The cylindrical part to be printed was called the keyring socket, and was divided into three parts, all cylindrical; as shown in Figure 19.

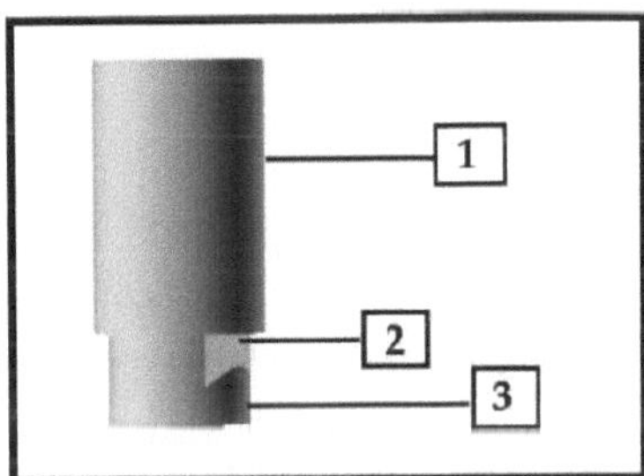

Figure 19: Representation of the keyring socket

Source: Prepared by the author

Part 1: this is the largest end of the piece, with a radius of 2.4 mm and a height of 10 mm. It will fit into one of the holes in face 3 of the die, whose radius is 2.45 mm (slightly larger than the radius of the end being built).

Part 2: represents the hole through which the ring will be placed, which is made up of a cylinder that runs perpondicular to part 3 of the keyring socket. The radius measures 1.25

mm.

Part 3: this is the smaller end of the socket, which will contain the hole through which the ring will fit. Its radius is 2.0 mm and its height is 5.0 mm.

Once the shapes and dimensions have been defined, the key ring slot must be programmed in Openscad so that it can be printed.

4.1.4. Object Ready and Print Results

After preparing the *Openscad* files for the dice and the key ring, they were converted into STL format (*Stereolithography*, the standard format for printing 3D models) and connected to the 3D printer to start printing. The codes for the printed objects can be found in Appendix A. The 3D printer used belongs to the GTMax3D Core A1 model and works with extrusion-based plastic, in which the extruder head releases subsequent layers of a heated plastic material in order to provide the desired contours for the object. The die and the key ring were printed separately, taking 31 minutes to print the die and approximately 11 minutes to print the key ring.

The material used was PLA (polylactic acid), which is a thermoplastic derived from renewable sources, such as corn starch, cassava roots or sugar cane; it is therefore a sustainable material (OLIVEIRA *et al.,* 2015). 0.21g of material was used to print the keyring insert and 8.95g to print the die. The average cost of the material used to produce a single object (dice and insert) was approximately R$ 0.97, R$ 0.94 for the dice and R$ 0.03 for the insert, resulting in a low cost, which makes it possible to sell the product at an affordable price and still generate a profit.

Below are some pictures of the object after it has been printed and is being used according to its possible functions. The product was tested and met all the specifications, satisfactorily performing the functions for which it was designed. Figure 20 shows a photo of the dice and insert after printing and Figure 21 shows a photo of the dice being used as a pencil holder.

Figure 20: Dice and keyring insert after being printed

Figure 21: Dice being used as lapel holders Source: Prepared by the author

It was also noted that the pencil holder function does not mean that only pencils can be placed on the object. It is also possible to support pens on face 1 (Figure 22), and other utensils, such as paintbrushes, can also be supported; in this case, as the tips are thinner, they should be supported in the holes on face 3, whose diameters are smaller (Figure 23).

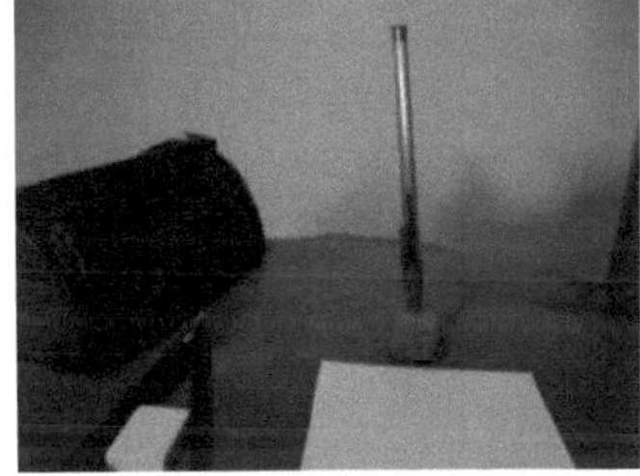

Figura 22:   Also used as a pen rest

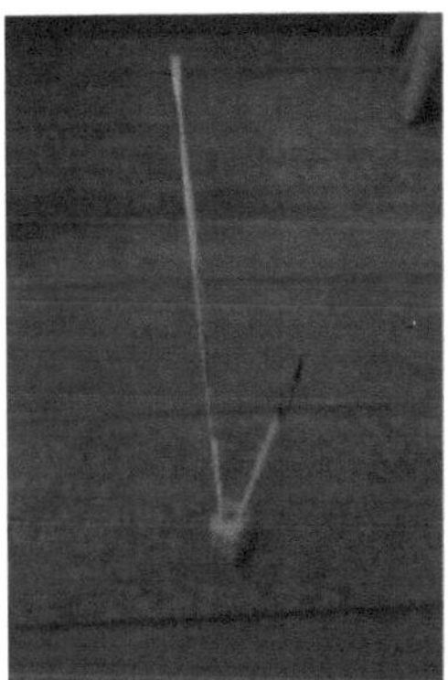

Figura 23:   Dice being used as a support for brushes

The object was also tested as a keyring (Figure 24), and as a decorative object in relation to a case (Figure 25).

Figure 24: Dice being used as a keyring Source: Prepared by the author

Figure 25: Dice being used as a decoration for a case

Source: Prepared by the author

4.2. 3D printing and creation of ergonomic profile object

Ergonomics is defined as a derivation of the Greek words ERGON (work) and NOMOS (rules), i.e. Ergonomics can be considered as the study of the laws of work (DUL and WEERDMEESTER, 1998).

Ergonomics applies information about human behavior, capabilities, limitations and characteristics **to the design of tools**, machines, tasks, jobs and environments for production, safe, comfortable and effective use (SANDERS; MCCORMICK, 1993).

Laville (1977) defines Ergonomics as a body of scientific and interdisciplinary knowledge relating to man and necessary for the design of instruments, machines and devices that can be used with maximum comfort, safety and efficiency.

3D printing can also be applied to ergonomics, by creating objects that help workers by reducing the physical strain caused by certain daily tasks. There are various types of these objects, known as ergonomic objects. The following are some brief examples:

•   Footrest: allows users to support their feet and thus rest their lower limbs. Figure 26 shows an example of a footrest:

Figure 26: Footrest Source: Portal Ergonomia do Trabalho

- Keyboard support: allows the user to support their wrists while using the keyboard, as shown in Figure 27:

Figure 27: Keyboard support Source: Portal Ergonomia do Trabalho

- Text holder: makes it easier to read documents, preventing the user from having to tilt their neck, which could cause discomfort or long-term injury. Figure 28 shows the use of a text holder:

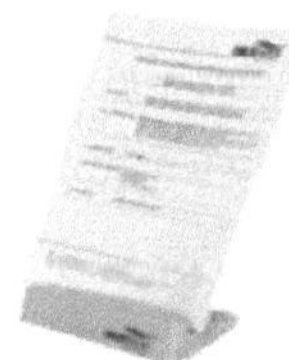

Figure 28: Text holder
Source: Portal Ergonomia do Trabalho

Several of these objects can be built with a 3D printer. The method also makes it possible to produce new or adaptable objects for a given situation. Another application will now be presented, again associated with the area of product planning, but this time linked to Ergonomics, showing that 3D printing can contribute to the creation of utensils that help workers carry out their activities in a more comfortable and satisfactory way.

This is a book rest that allows the user to place the book against the support so that they can read without having to tilt their neck. Frequent tilting of the neck can cause injuries, pain and discomfort in this region, resulting in reduced productivity. By using the book rest, users can read more easily and comfortably, as well as extending the duration of their activity without damaging their necks.

To make this product, two attempts were necessary, as the first attempt did not achieve the planned results.

4.2.1. First attempt

In the first attempt, the object was planned according to figure 29 and divided into four parts:

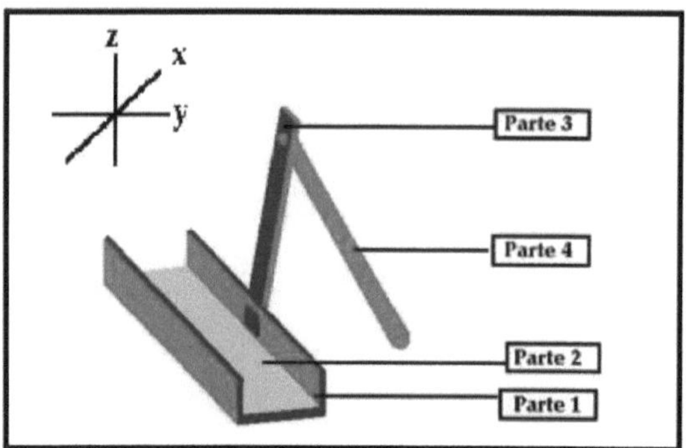

Figure 29: Planned shape of the bookend Source: Prepared by the author

Part 1: Channel: this is the base of the support and consists of a parallelepiped with dimensions of 300 mm long, 62 mm wide and 32 mm high, from which another parallelepiped defined in part 2 must be removed.

Part 2: Removed parallelepiped: which is subtracted from the base (part 1) to create the channel into which an open book will fit. The measurements of this parallelepiped were designed precisely for this purpose, with a length of 320 mm (a little longer than the length of part 1, so that it would be open on both sides to fit longer books), a width of 54 mm, and a height of 26 mm; all these measurements were designed so that the open book could be fitted in satisfactorily.

Part 3: Backrest: this is a parallelepiped with dimensions of 300 mm long, 5 mm wide and 150 mm high, whose purpose is to support the back of the book. It is rotated -10° in relation to the X axis (in Openscad), in order to generate a slight inclination that allows the book to be supported ergonomically.

Part 4: Backrest support shaft: this is a cylinder whose function is to ensure the balance of the book rest, inspired by the support of a picture frame. Its dimensions are 6 mm for the radius and 162 mm for the height. It is rotated 20° in relation to the x-axis, 20° in relation to the y-axis and -43.5° in relation to the z-axis, so that it can balance the backrest more precisely.

After inserting the book support code, an image is generated on the Openscad screen that depicts the planned object (Figure 30). Extra code was also added in an attempt to generate an image of an open book (depicted in transparent and not part of the object), which aims to demonstrate the desired shape for the book to be supported.

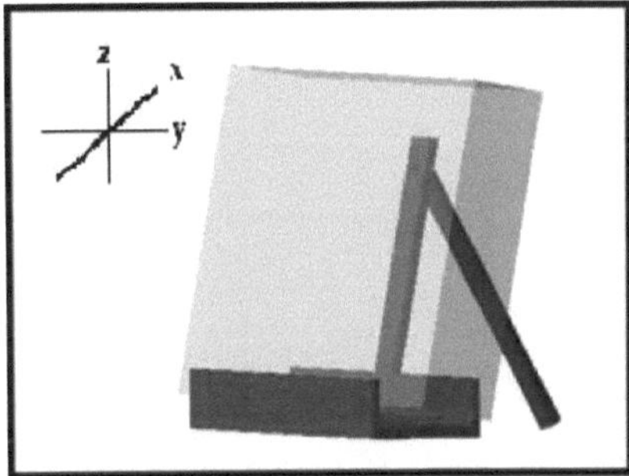

Figure 30: Image of the bookend generated by Openscad and its planned use
Source: Prepared by the author

The object was printed using PLA, with an estimated printing time of 8 hours and 6 minutes. 223.54 g of material were consumed, and the cost of producing the bookend was approximately R$26.83. Again, this was a low cost, which means that it is an affordable product.

The image of the object after printing is shown in Figure 31:

Figure 31: Support for printed book (first attempt) Source: Prepared by the author

After printing, tests were carried out. Unfortunately, the object didn't work quite as planned. This was due to the fact that the widths of parallelepipeds 1 and 2 were too small, so only thinner books were able to stand correctly. Thicker books (with more pages) didn't lean. This means that the width of part 2, whose value was 54 mm, was too small, so the value of the width of the shank, and therefore part 2, should also be increased.

Below are photos of the tests carried out on the bookend.

Figure 32: Bookend being tested on a thick book Source: Prepared by the author

As can be seen in Figure 32, although the open book was supported, this did not happen as planned, as there was no inclination designed for the book, as shown in Figure 30, which makes reading difficult.

On the other hand, it's worth noting that the support worked exactly as planned in the case of sheets of paper and thinner books, such as a diary, as shown in Figures 33 and 34 below.

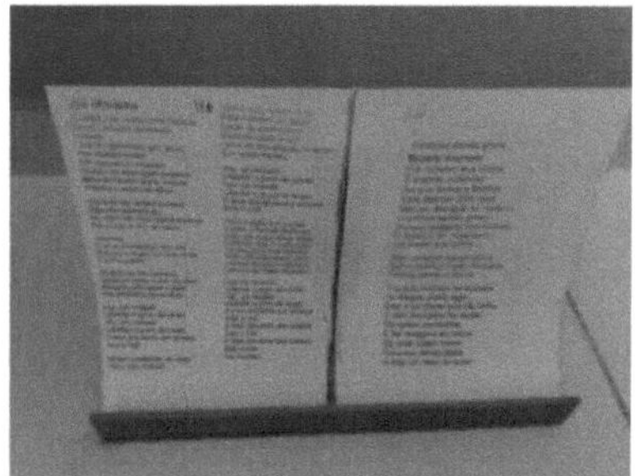

Figure 33: Bookend being used for a sheet of paper Source: Prepared by the author

Figure 34: Bookend being used for a diary Source: Prepared by the author

4.2.2. Second attempt

As the first attempt did not achieve the planned results, it was necessary to make a second attempt. As well as thinking of a way to fix the problems presented in the first attempt, we also thought of a way to improve the functioning of the product: to make a collapsible bookend, which would make it much easier to transport. To this end, the object was designed as shown in Figure 35, divided into four parts:

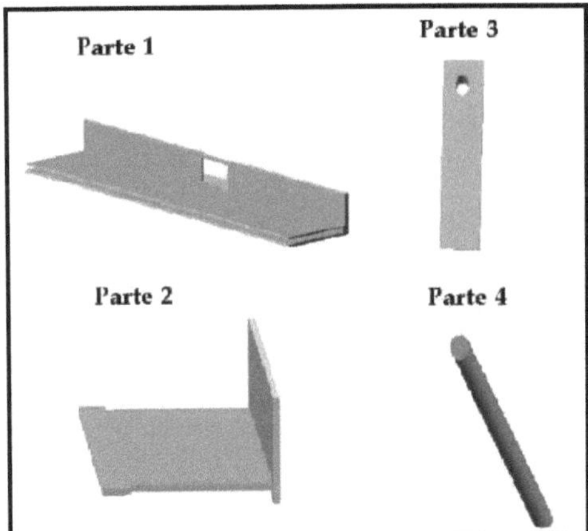

Figure 35: Parts that make up the second attempt at the bookend
Source: Prepared by the author

Part 1: Extendable channel: formed by the difference between a parallelepiped whose dimensions are 300 mm long, 62 mm wide and 32 mm high, and another cube, whose dimensions are 320 mm high, 58 mm wide and 25 mm high. The purpose of this channel is to act as a kind of "drawer", containing a double floor in its lower part with three windows, allowing part 2 to be fitted and the base to be adjustable. The dimensions of the windows that will allow part 2 to be fitted are 320 mm long, 54 mm wide and 4 mm high. It will also contain a gap with dimensions of 30.6 mm long, 5 mm wide and 20 mm high, which will allow part 3 to be fitted. Finally, there will be two small parallelepipeds at the ends of the double floor, with dimensions of 9 mm long, 5 mm wide and 6 mm high, forming the windows, which will allow part 2 to be fitted as a limit to how far it can extend.

Part 2: Extendable channel: this acts as an extension to the base of the support and is constructed by joining three parallelepipeds. The first is 274 mm long, 44 mm wide and 2 mm high. The second is 294 mm long, 8 mm wide and 2 mm high. The second parallelepiped is slightly longer than the first, precisely so that it can be braked by the two smaller parallelepipeds installed on the double floor of part 1, when it is pulled to its maximum length. This part will be fitted into the longer window at the base of part 1.

Part 3: Backrest: this is a parallelepiped measuring 30 mm long, 5 mm wide and 150 mm high. It has a hole in its upper part, which corresponds to an elliptical cylinder with a radius of 5.1 mm and a height of 160 mm (through which part 4 will be fitted). Its purpose is to allow the open book to be supported and tilted at a convenient angle for reading.

Part 4: Backrest support shaft: a cylinder with a radius of 4.5 mm and a height of 160 mm, which will be fitted into part 3, allowing a favorable inclination to support the book.

After modeling in Openscad, the object was printed, with each part being printed separately, unlike the first attempt in which the object was printed all at once. The costs and times taken to print each part are:

- Part 1: printing time of 4 hours and 42 minutes, 86.61 g of material and a cost of R$ 10.39.

- Part 2: printing time of 2 hours and 43 minutes, 48.80 g and cost of R$ 5.86.

- Part 3: time spent 52 minutes, 12.92 g and cost of R$1.55.

- Part 4: time spent of 38 minutes, 8.97 g and cost of R$1.08.

As a result, the total time taken to print the complete support was approximately 8 hours and 55 minutes and the total cost was R$18.88, which was even less than the first attempt. This was due to the fact that the parts were printed separately and had more hollow spaces.

Figure 36 shows an image of the four parts after they have been printed:

Figure 36: The four parts of the second attempt, immediately after printing

Source: Prepared by the author

The bookend was then assembled as planned, and Figure 37 shows how the finished object looks:

Figure 37: Assembled bookend

Source: Prepared by the author

As the base is self-adjusting, it can be extended to support heavier books, as shown in Figure 38.

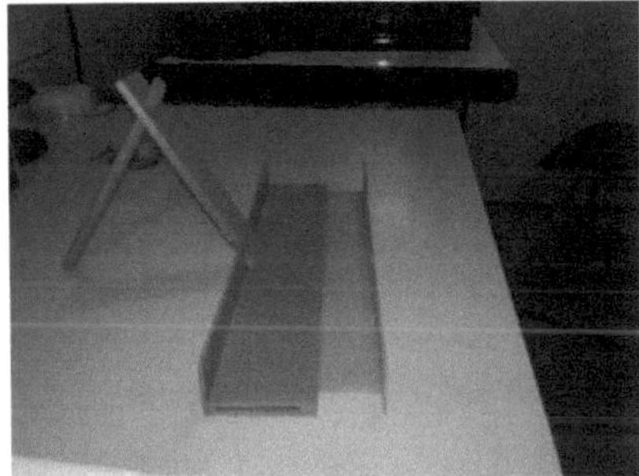

Figure 38: Book rest base extended

Source: Prepared by the author

Tests were carried out for the second attempt at the bookend, as shown in Figure 39.

Figure 39: Test for the second attempt at the bookend

Source: Prepared by the author

Thus, it can be seen that on the second attempt the object worked as expected. However, it is important to note that it still has some limitations. The first of these is that you can only place books whose width or back is no more than two centimeters. If it's wider than that, the book won't stay open and the support won't support its weight. Another restriction is the height of the base, which has been reduced, making it impossible to support more closed books.

Despite its limitations, the support constructed can be used satisfactorily for light, thin books, for example magazines, as illustrated in Figure 40:

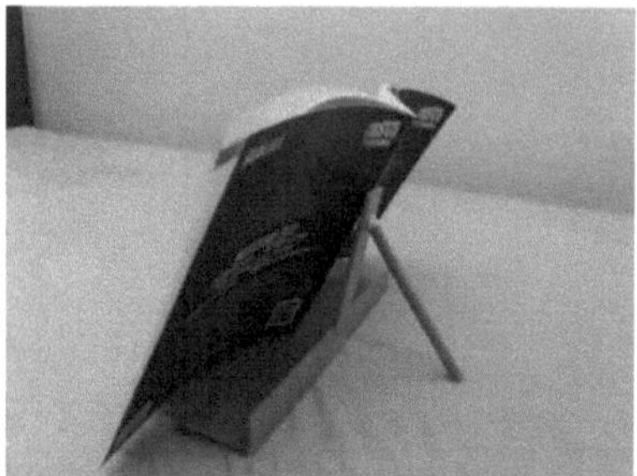

Figure 40: Bookend being tested for a magazine, seen from behind Source: Prepared by the author

In order for it to function more adequately and more comprehensively, and to be able to support thicker or heavier books, the object needs to undergo a series of improvements. For example, we can mention the addition of grids on the base, so that it can be more stable when extended. We could also mention increasing the radius of the support cylinder, so that it can support heavier books, and increasing the height of the base, so that thicker and more closed books can be supported. It is worth noting that in order to better fit part 3 (backrest) into part 1 (cane), it was necessary to drill a hole mechanically, as this had not been planned previously. Therefore, this hole should already be planned when designing a new model.

It can be concluded that the product is satisfactory for use with thin books, magazines or sheets of paper. However, it needs to be improved so that it can be used with more options, such as thicker books.

Finally, a comparison can be made between the amounts of cost, time and material spent on the first and second attempts to build the bookend. The comparison is shown in Table 6:

Table 6: Comparison of attempts to develop support

Bookends		

	Costs	Material	Time
First attempt	R$ 26,83	223,54 g	8 h and 6 min
Second attempt	R$ 18,88	157,3 g	8 h and 55 min

As you can see, the second attempt was more economically viable than the first, both in terms of costs and materials. Although it took longer to print, due to the fact that the parts were printed separately, it can be concluded that the second attempt produced better results than the first, both in terms of cost and practicality, by providing a detachable bookend.

CHAPTER 5

FINAL CONSIDERATIONS

5.1. Conclusion

In this work, a general analysis was made of 3D printing, discussing various aspects of this technology, such as possible fields of application, advantages, disadvantages, growth prospects, history, evolution, etc. In addition, a brief analysis was made of its operation and assembly, as well as the advantages and disadvantages of buying a ready-made or unassembled 3D printer.

Based on everything that has been presented, we can conclude on the great importance of 3D printing and its possible applications in various areas. Among these, its use in relation to product planning was emphasized, showing how the method can contribute, through the construction of prototypes, or even through the creation of new types of products, enabling the user to put their creativity into practice and design objects for their leisure, personal needs, or invest in creations aimed at making a profit. To this end, two practical applications were carried out, producing two different products using 3D printing.

The first application involved the planning, design and production, using 3D printing, of a product consisting of a dice that could also be used as a keyring, pencil holder and decorative object. The object created was considered to be innovative and attractive in relation to its target audience, due to the fact that it had several possible forms of use, which set it apart from other similar products. After the object had been printed according to its specifications, it was possible to carry out tests, thus verifying that it successfully fulfills the functions for which it was designed, with minimal material and time costs.

In the second application, the entire planning and production process for a bookend was presented, a product that falls within the area of

Ergonomics, as it allows the user to read more easily and comfortably. The bookend was designed to consume as little material as possible for its production, containing only the essential parts to successfully fulfill its function. Two attempts were made at this application. In the first attempt, the object only worked for very thin sheets or books, such as diaries. To try to solve this problem, another attempt was made. After being printed for the second time, tests were carried out which proved that the object satisfactorily fulfills its function in relation to some types of books, saving costs and materials, but it still had restrictions in relation to thick and heavy books, as these compromise the balance of the support. Therefore, based on the prototype created after the second attempt, improvements were proposed for the design of a new object that can be used in a more comprehensive and satisfactory way.

Spatial geometry was used to build both models, using parallelepipeds and cylinders to form the objects. It can therefore be concluded that 3D printing can contribute to the development of the most diverse types of products, through the combination of various existing geometric shapes, the basis of projects and the development of products in general.

In addition to the applications shown here, there are a multitude of products that can be created with the same purpose as the one presented here, or with various other purposes, leaving it up to the user to choose how they will take advantage of this fantastic technology, which over time tends to be available to more and more people.

5.2. Suggestions for future work

After all that has been done and discussed in this paper, we can see the great potential of 3D printing. However, there are still some points that need to be improved in order to enhance the use of this promising technology. For this reason, future work could include research aimed at optimizing 3D printing, in an attempt to find ways of minimizing waste, such as the loss of material that occurs when light falls and production has to be restarted, or minimizing the losses that occur when printing floating parts.

Another suggestion for future work could be to make a third attempt at designing the bookend, in an attempt to improve the problems that were still present even after the second attempt.

BIBLIOGRAPHICAL REFERENCES

AGUIAR, L. C. D. **A process for using 3D printing technology in the construction of teaching aids for science teaching.** Dissertation (Master's in Science Education) - Faculty of Sciences, UNESP, Bauru, 2016.

BAIÀO, F. J. **3D printer functionalities and technologies.** Dissertation (Degree in Computer Engineering) - Universidade Sâo Francisco, Itatiba, 2012.

BAIXAKI, **Blender: one of the most acclaimed 3D modeling and animation softwares gains new features in its new update.** Available at: <http://www.baixaki.com.br/download/blender.htm>. Accessed on June 7, 2016.

BAIXAKI. **Autocad: Bring your technical projects to life using the latest version of the most famous CAD software on the market.** Available at: <http://www.baixaki.com.br/download/autocad.htm>. Accessed on June 8, 2016.

BAXTER, M. **Projeto de produto: Guia pràtica para o design de novos produtos.** 2ª ed. Sâo Paulo: Editora Blucher, 2000.

BELL, C. **Maintaining and Troubleshooting your 3D Printer.** New York: Apress, ia Ed.; 2014, 498 p

BENVINDO, A. Z. **Naming in the process of constructing the waste picker as an economic and social actor.** 2010. Dissertation (Master's Degree) - University of Brasilia, Brasilia, 2010.

BOURELL, L. D.; BEAMAN, J. J.; LEU, C. M.; ROSEN, W. D.**A Brief History of Additive Manufacturing and the 2009 Roadmap for Additive Manufacturing: Looking Back and Looking Ahead.** Rapid Tech 2009: US-TURKEY Workshop on Rapid Technologies, Istanbul, September 24, 2009

CAMPBELL, I.; BOURELL, D.; GIBSON, I. **Additive Manufacturing: rapid prototyping comes of age.** Rapid Prototyping Journal, v.18, n.4, p.225-228, 2012

CHIAVENATO, Idalberto. **People management.** Rio de Janeiro: Elsevier, 2005. COSTA, C.S. da; MIGUEL, D. C. V; FILHO, E. S.; SILVA, M. V. F.; MACHADO, M. C. **3D printing: The technology of the moment.** Revista Pós em Revista, Issue 11, 2015.

DUL, J.; WEERDMEESTER, B. **Practical ergonomics.** Sao Paulo: Edgard Blucher, 1998. 147 p.

EPOCA NEGOCIOS. **Industry resorts more to 3D printing, and use of the technology** grows by 30%. Available at: <http://epocanegocios.globo.com/Economia/noticia/2017/02/industria-recorre-mais-impressao-3d-e-uso-da-tecnologia-cresce-30.html>. Accessed on May 3, 2017.

EVANS, B. **Practical 3D Printer.** New York: Apress, 2012. 206 p.

FERNANDES, F. F. **Entrepreneur profile and its influences on local development: a study focusing on the use of technologies.** Postgraduate Program in Regional Development, Centro Universitàrio Municipal de Franca, Sao Paulo, 2016.

FLORÊNCIO, E. Q., FERREIRA, D. B. F., QUINTELLA, I. P. C. P. **The future of the construction process? 3D printing in concrete and its impact on the design and production of architecture.** XX Congreso de la Sociedad Ibero-americana de Gràfica Digital 9-11, November, 2016 - Buenos Aires, Argentina.

FREIXO, Osvaldo M., TOLEDO, J. C. de. **Product Life Cycle Cost Management during the Development Process.** IV Congr. Bras. Gestao e Desenv. de Produtos - Gramado, RS, Brazil, October 6-8, 2003.

GARCIA, L.H. T. **Development and manufacture of a mini 3D printer for ceramics.** Dissertation from the Postgraduate Program in Mechanical Engineering in the Area of Concentration in Mechanical Design. Sao Carlos, 2010.

GLOBO.COM, 2015. **First car made with a 3D printer in China drives through the country's streets.** Available at:

<http://g1.globo.com/tecnologia/noticia/2015/03/primeiro-carro-feito-com-impressora-3d-roda-pelas-ruas-na-china.html>. Accessed on June 2, 2016.

IBGE - BRAZILIAN INSTITUTE OF GEOGRAPHY AND STATISTICS. 2010 Demographic Census. Rio de Janeiro: IBGE, 2012a.

3D printing made easy. **Learn about the different types of materials for 3D FDM printing.** Available at: <http://www.impressao3dfacil.com.br/conheca-os- different-types-of-materials-for-3d-fdm-printing/>. Accessed on April 27, 2017.

INFORÇATTI NETO, P. **Study of the technical feasibility and design of a mini screw extrusion head for portable three-dimensional printers.**2013. 127p. Dissertation (master's degree) - Sao Carlos School of Engineering, University of Sao Paulo, Sao Carlos, 2013

JAI BERT, F.; BOETTO, S.; NADON, F.; LAUWERS, F.; SCHIMIDT, E. & LOPEZ, R. **One-step primary reconstruction for complex craniofacial resection with PEEK custom-**

made implants. J. Cranio maxillo fac. Surg., 42(2):141-8, 2014.

JUNIOR, P. O. C., MARQUES, D. M. N. **3D printers: reducing cost and time in product development**. Industrial Mechatronics Technology, Garça Technology College, 2013.

KHOSHNEVIS, B. **Contour Crafting: Automated Construction**: BehrokhKhoshnevisatTEDxOjai, 2012. Available at: <https://www.youtube.com/watch?v=JdbJP8Gxqog>.

LAVILLE, A. **Ergonomics**. Translated by Màrcia Maria Neves Teixeira. Sao Paulo: EPU-University of Sao Paulo, 1977. 99p.

LEMOS, Manoel. Blog Fazedores. [2013; 2014]. Available at: <http://blog.fazedores.com/>. Accessed on: 25 Apr. 2014

LIMA, T. S., CORRÊA, G. R. **influence of new technologies for designers and markers**. Brazilian Congress of Research and Development in Design, Belo Horizonte 04 to 07 Oct. 2016.

MAGALHÂES, Beatriz J. **Liminaridade e exclusâo: os catadores de materiais reciclâveis e suas relações com a sociedade brasileira**. 2012. Dissertation (Master's Degree) - Federal University of Minas Gerais (UFMG), Belo Horizonte, 2012.

MARIOTTO, A. F. (2013). **Study and improvement project for a 3D printing machine**. University of São Paulo, Sâo Carlos School of Engineering.

MONTEIRO, F. T. M. **3D printing in the productive environment and design: a study in jewelry manufacturing**. Dissertation (Postgraduate Program in Design, Belo Horizonte, State University of Minas Gerais, 2015).

MWC, 2016. **How about making pizza for dinner?** Available at: <http://brasil.elpais.com/brasil/2016/02/25/tecnologia/1456394796_311293.html>.

Accessed on June 2, 2016.

MORAES, P. H., OLATE, S., CANTiN, M., ASSIS, A. F., SANTOS, E., SILVA, F. O., SILVA, L. O, 2015. **Anatomical Reproducibility through 3D Printing in Cranio- Maxillo - Facial Defects**. Available at:

<http://www.scielo.cl/scielo.php?script=sci_arttext&pid=S0717-95022015000300003&lang=pt>. Accessed on July 27, 2016.

NETO, A. S. **Discover why Agile Engineering is the future of Product Design**, Available at <http://blog.wishbox.net.br/2016/06/10/engenharia-agil- futuro-do-design-de-

produto/>. Accessed on May 24, 2017.

OLIVEIRA, H. R. F.; MORALES, G.; ZEFERINO. L. H. **Cast Material Deposition Modeling - Introduction and Practice Applied to Production Engineering.** XXII SIMPEP, Baurù, SP., 2015.

PALLAROLAS, E. A. F. F. **Technical review of additive manufacturing processes and study of configurations for three-dimensional printer structures.** University of São Paulo, São Carlos School of Engineering, 2013.

WORK ERGONOMICS PORTAL. **Ergonomic products.** Available at: <http://www.ergonomianotrabalho.com.br/produtos.html>. Accessed on April 21, 2017.

PANKIEWCZ, I. **How does the 3D printer work?** 2009. Available at: <https://www.tecmundo.com.br/impressora/2501-como-funciona-a-impressora-3d- .htm>. Accessed on: September 10, 2016.

REPRAP.PT, **RepRap - 3D Printers**. Available at: <http://www.reprap.pt/>. Accessed on 03 Dec. 2016.

REPRAP WIKI, **RepRap/en**. Available at: < http://reprap.org/wiki/RepRap/pt>. Accessed on 04 Dec. 2016.

RIBEIROS, A., MANOEL, C. E., SILVA, M. G., DELGADO NETO, G. G., **Planning a methodological proposal for analyzing the quality of products produced in 3D printers.** Intellectus Magazine, Year IX, N° 23, 2013.

ROZENFELD, H.; AMARAL, D. C. FORCELLINI, F. A.; AMARAL, D. C.; TOLEFO, J. C. SILVA, S. L.; ALLIPRANDINI, D. H.; SCALICE, R. K. **Gestao de Desenvolvimento de Produtos: uma referência para a melhoria do processo.** p.542. Sao Paulo: Saraiva, 2006.

SANDERS, M. S.; MCCORMICK, E. J. **Human factors engineering and design.** McGraw-Hill: New York, 1993.

SIGN SILK and SCREEN, 2015. **3D printing on the list of trends for 2016.**
Available at: <http://signsilk.com.br/impressao-3d-e-tendencias-tecnologicas-2016/>. Accessed on June 9, 2016.

SILVA, P. L.; GOES L. F.; ALVAREZ, R. A. **Current situation of recyclable and reusable material collectors** - Brazil. IPEA - Institute for Applied Economic Research, Brasilia, 2013.

SOFTPEDIA. **Create and compile solid 3D CAD objects and view their graphical representation, while working in a user-friendly and intuitive environment**. Available at:<http://www.softpedia.com/get/Science-

CAD/OpenSCAD.shtml>. Accessed on June 9, 2016

SOUZA, A. F. and ULBRICH, C. B. L. **Computer Integrated Engineering and CAD/CAM/CNC Systems - Principles and Applications**. Sao Paulo: Artliber Editota, 2009.

SOUZA, B. de A. 2016. **3D printers: the future of crime?** Criminal Sciences Channel. Available at:

<http://canalcienciascriminais.jusbrasil.com.br/artigos/296840886/impressoras-3d-o-future-of-criminality>. Accessed on October 27, 2016.

TAKAGAKI, L. K. CHAPTER 3. **3D Printing Technology**. Revista Inovaçâo Tecnològica, Sao Paulo, v. 2, n. 2, p. 28 - 40, dec 2012. ISSN 21792895.

TECMUNDO, 2013. **13 3D printers already available in Brazil**. Available at: <http://www.tecmundo.com.br/impressora-3d/40248-13-impressoras-3d-ja- disponiveis-no-brasil.htm>. Accessed on June 8, 2016.

TECMUNDO, 2013. **20 questions and answers about 3D printers.** Available at: <http://www.tecmundo.com.br/impressora-3d/39647-20-perguntas-e-respostas-   sobre-impressoras-3d.htm>. Accessed on June 3, 2016.

TECMUNDO, 2015. **First Chinese car made by a 3D printer takes to the streets**. Available at: <http://www.tecmundo.com.br/carro/77401-primeiro- carro-chines-fabricado-impressora-3d-ruas.htm>. Accessed June 3, 2016.

TERRA, 2012. **3D printing: meet the machines that will revolutionize the world**. Available at: <http://tecnologia.terra.com.br/eletronicos/impressao-3d-   conhecera-as-maquinas-vao-revolucionar-omundo,ca1afbd1680ea310VgnCLD200000bbcceb0aRCRD.html>. Accessed on June 13, 2016.

VASCONCELOS, P. **O Fabrico Ràpido de Ferramentas ao Serviço da Engenharia Concorrente**. Tecnometal Publication, 2001. Polytechnic Institute of Viana do Castelo, Viana do Castelo, Portugal.

VOLPATO, N.; SILVA, J. V. L; SANTOS, J. R. L; CARVALHO, J; PETRUSH, G; FERREIRA, C. V; AHRENS, C. H. **Rapid prototyping - technologies and applications**.

Sao Paulo: Edgar Blücher, 2007.

WIKI, 2016. RepRap/en. Available at: <http://reprap.org/wiki/RepRap/pt>. Accessed on 04 Dec. 2016.

WOHLERS, T. T. **Wohlers Report**. 2008.

APPENDIX A - OPENSCAD OBJECT CODES

Code for data preparation:

```
size=24;

angles=2.6;

artefact=0.05;

moduledice_empty() {

minkowski() {

internal_size=(8-angles)*size/8;

external_size=angles*size/8;

cube([internal_size,internal_size,internal_size]);

translate([external_size/2,external_size/2,external_size/2]) sphere(d=external_size,$fn=100);

}}

module hole(hsize=1) {

hole_width=[7.7, 6.8, 4.9, 5.2, 4.8, 4.7];

hole_height=[8, 5.1, 6.6,4.4,4.1, 3.6];

cylinder(d=hole_width[hsize-1],h=hole_height[hsize-1],$fn=100);

}

difference() {

color("green")dice_empty();

color("orange"){

rotate ([180, 0, 0])

translate([size/2,-size/2,-size-artefact])

translate([0,0,0]) hole(1);

rotate ([0, 0, 0])

translate([size/3,size/4,-artefact]) {

translate ([0,0,0]) hole(6);

translate([0,size/4,0]) hole(6);

translate([0,size/2,0]) hole(6);

translate([size/3,0,0]) hole(6);
```

```
translate([size/3,size/4,0]) hole(6);

translate([size/3,size/2,0]) hole(6);

}

rotate ([90, 0, 180])

translate([-size/3,size/3,-artefact]) {

translate([0,size/3,0]) hole(2);

translate([-size/3,0,0]) hole(2);

}

rotate ([90, 0, 0])

translate([size/3,size/3,-size-artefact]) {

translate([0,0,0]) hole(5);

translate([0,size/3,0])     hole(5);

translate([size/3,0,0])     hole(5);

translate([size/3,size/3,0]) hole(5);

translate([size/6,size/6,0]) hole(5);

}

rotate ([0, 90, 180])

translate([-size/3,-size/3,-size-artefact]) {

translate([0,0,0]) hole(3);

translate([-size/3,-size/3,0]) hole(3);

translate([-size/6,-size/6,0]) hole(3);

}

rotate ([0, 90, 0])

translate([-size/3,size/3,-artefact]) {

translate([0,0,0]) hole(4);

translate([-size/3,size/3,0]) hole(4);

translate([-size/3,0,0])    hole(4);

translate([0,size/3,0])    hole(4);

}}}
```

Code for making the keyring socket

```
cylinder(r=2.4,h=10, $fn=100);

union ()

difference() {

cylinder (h = 5, r=2, center = true, $fn=100);

rotate ([90,0,0])

cylinder (h = 2.4, r=2, center = true, $fn=100);

}
```

Code for book support (first attempt),

```
//color("green")

union(){{

difference () {

cube ([300, 62, 32],center=true);

translate ([0,0,4])

cube ([380, 54, 26], center=true);

}

union () {

difference(){

union(){

translate ([0,123,-13])

rotate ([20,20,-43.5])

cylinder (r=6, h=392, $fn=150,center=true);

color("red")rotate([-10,0,0])

translate ([-15,23,0])

cube ([30,5,150]);

}

rotate([-10,0,0])translate ([0,3,130])cube ([600, 40, 325],center=true);

translate ([0,0,5])

cube ([380, 54, 30], center=true);
```

```
color("blue")translate ([0,50,-166])

cube ([380, 354, 300], center=true);

}}}}
```

Code for book support (second attempt):

Parts 1 and 2:

```
union(){{

difference () {

cube ([300, 62, 32],center=true);

translate ([0,0,4])

cube ([320, 58, 25], center=true);

translate ([0,0,-12])

cube ([340, 54, 4], center = true);

translate ([0, -28, 0])

cube ([320, 3, 40], center=true);

rotate([-13,0,0])

translate ([-15,25,0])

cube ([30.6,5,150]);

}

union () {

translate ([-138, -30, -13])

cube ([274, 44, 2]);

translate ([-147,13,-13])

cube ([294,8,2]);

color ("green")

translate ([-150, -23,-16])

cube ([9,5,6]);

color ("green")
```

```openscad
translate ([141,-24,-16])

cube ([9,5,6]);

}}}
```

Part 3:

```openscad
difference ()

{

rotate([-13,0,0])

translate ([-15,25,0])

cube ([30,5,150]);

translate ([0,125,-13])

rotate ([20,20,-43.5])

cylinder (r=5.1, h=160, $fn=150);

}
```

Part 4:

```openscad
union () {

difference(){

union(){

translate ([0,125,-13])

rotate ([20,20,-43.5])

cylinder (r=4.8, h=160, $fn=150, center= true);

}

translate ([0,50,-234])

cube ([380, 354, 300], center=true);

rotate([-13,0,0])

translate ([0,54,130])

cube ([600, 40, 325],center=true);

}}
```

yes I want morebooks!

Buy your books fast and straightforward online - at one of world's fastest growing online book stores! Environmentally sound due to Print-on-Demand technologies.

Buy your books online at
www.morebooks.shop

Kaufen Sie Ihre Bücher schnell und unkompliziert online – auf einer der am schnellsten wachsenden Buchhandelsplattformen weltweit! Dank Print-On-Demand umwelt- und ressourcenschonend produziert.

Bücher schneller online kaufen
www.morebooks.shop

Printed by Books on Demand GmbH, Norderstedt / Germany